SQUASH

A COUNTRY GARDEN COOKBOOK

SQUASH

A COUNTRY GARDEN COOKBOOK

By Regina Schrambling

Photography by Deborah Jones

CollinsPublishersSanFrancisco

A Division of HarperCollinsPublishers

For Bob, who brought me back to food.

First published in USA 1994 by Collins Publishers San Francisco
Copyright © 1994 Collins Publishers San Francisco
Recipes and text copyright © 1994 Regina Schrambling
Photographs copyright © 1994 Deborah Jones
Food Stylist: Sandra Cook
Prop Stylist: Sara Slavin
Creative Director: Jennifer Barry
Designer: Kari Perin
Series Editor: Meesha Halm
Library of Congress Cataloging-in-Publication Data
Schrambling, Regina
Squash: a country garden cookbook / by Regina Schrambling:
photography by Deborah Jones.
p. cm.
Includes index.
ISBN 0-00-255346-5
1. Cookery (Squash) I. Title.
TX803.S67S37 1994
641.6'562--dc20 CIP 94-2112

Acknowledgments
Collins and the photography team would like to thank
Jeri Jones and Helga Sigvaldadottir, photo assistants; Allyson Levy
and Kathleen Fazio, food styling assistants; Kristen Wurz,
design and production coordinator; and Jonathan Mills, production
manager. Thanks also to Bess Petlak at Freida's Inc.;
John Gantner and Nancy Walker of School House Vineyards in
St. Helena; Molino Creek Farm Collective in Davenport;
Allison Evans; Missy Hamilton; and Joan Hertzberg.

CONTENTS

Introduction 7

Glossary 9

Openers 19

Accompaniments 39

Main Courses 57

Sweets 75

Metric Conversions 94

Index 95

INTRODUCTION

I came late to the squash fan club. Although my childhood was spent in the Southwest, where some of these versatile vegetables originated, I don't remember eating more than pumpkin on a regular basis. My mother did her vegetable gardening in the canned goods aisle at the local grocery, and pumpkin was a staple only because all our neighbors were Mexicans who didn't wait around for Thanksgiving to eat it. They baked it into sweet *empanadas* all year. They savored the seeds, roasted and salted, as *pepitas*. And they even taught us to eat the blossoms off the vine, battered and deep-fried.

But it was not until I ripened into a professional eater in New York City that my own appetite for squash truly bloomed. Partly it was piqued by exposure, since so many restaurants—ethnic and American—showcase squash in everything from soup to tarts, from risotto to enchiladas. But it was also stimulated by availability: Any produce stand now routinely stocks a minimum of six to eight varieties, in every shape and color.

When I went off a decade ago to train as a chef, I had never tasted even a squash as mundane as butternut. My addiction started during a class about vegetables when Stephanie, the one student more interested in restaurant management than cooking, produced what she justifiably boasted was the best dish of the day. It was nothing more than a simple purée of butternut with a bit of honey, a little butter and fresh thyme, but it was simply spectacular.

The lesson from then on has been that great squash has little to do with the cook and everything to do with the ingredient. No expertise is needed to cut a delicata in half and bake it. The flavor stands alone.

Squash also comes in so many varieties that a cook can shine for weeks producing different dishes using essentially the same ingredient. Most varieties are sold year-round, but this remains one vegetable guaranteed to keep us aware of the seasons. Summer is high time for crooknecks and sunbursts and cymlings, not to mention squash blossoms and baby squash. In fall, when the zucchini are swelling to blimp size, the first winter varieties roll off the vines: pumpkin and turban, buttercup and Hokkaido. And even in darkest winter, when potatoes and onions are the main staples, there is always some kind of squash available to brighten up both markets and menus.

The population explosion in the squash cornucopia is partly due to a new realization that this varied vegetable doesn't just taste good, it is also one of the best choices a health-conscious eater can make. Winter squash in particular are extremely high in beta carotene, the antioxidant that has been credited with reducing the risk of everything from common ailments to cancer. All squash are also low in calories, high in other vitamins and minerals and full of fiber. And at a time when nutritionists are advocating eating five portions of fruits and vegetables daily, there's a squash for each serving, from muffins to main dishes.

Squash has been a vital ingredient in North American kitchens for literally centuries. Along with beans and corn, it formed the holy trinity of the native diet long before Columbus set sail. When the conquistadors arrived in the Southwest in the early 1500s, they were taught by Native Americans to cook with every part of squash, including the seeds and the flesh, which they dried on stakes in the sun to ensure provisions for winter.

The name *squash* actually comes from the Narragansett Indian word *askútasquash*, meaning "a green thing eaten raw," which sounds like the worst way to consume it. Once the Pilgrims came along, they adapted squash to their diets and squash found a place on the fire.

Thanks to this New World bounty, cuisines all over the world, from Italy to India, have been enriched. Because of my background I'm most inclined to give squash a Mexican accent. But since I now live in New York City, the ultimate melting pot, the recipes in this book showcase more universal flavors.

And all of them reflect my late-blooming fondness for squash. My consort often seems baffled when he hears me gushing over a newfound variety, savoring it the way some people do truffles or foie gras and insisting he agree on its wonders. "I like squash fine," he'll say, "but you *love* it." Converts are always the most devoted, especially when it comes to squash.

GLOSSARY

Squash is a many-splendored food. Unlike other vegetables, it's available all year, in two varieties, summer and winter (names that ironically reflect the last vestige of seasons in the world of produce). It also comes in a rainbow of colors, from earth tones to deep green to pale blue, and in a subtle but rich range of flavors. And while it is one of the oldest foods cultivated by North Americans, it's a category wide open to innovation. Botanists are constantly creating crossover dream hybrids, with sweeter flesh and more vivid shells.

Zucchini, of course, is easily the most popular, prolific and versatile squash of all time, and its dominance is reflected in the recipes that follow in this book. But other squash are no less worthy of serious exploration.

The varieties listed in this book are those most widely available in either supermarkets, farmers' markets or health-food stores; the latter are particularly good sources for new types from organic producers. Gardeners who grow their own, of course, have access to far more.

Selecting: With winter squash, look for smooth skins with no cuts or soft spots. The centers are mostly seeds, so choose squash that feel heavy for their size. Some of the more gargantuan varieties such as calabaza and Hubbard are often sold in chunks, seeded and wrapped in plastic film; these should look moist, not desiccated.

With summer squash, choose those that are firm to the touch. The skin should be smooth, with no bruises or withered areas. Smaller squash are generally better for most dishes because they have less of the soft center to overcook and also have a more delicate flavor. Oversized zucchini are good principally for soups or for stuffing.

Cleaning: Summer squash fresh from the field or garden should be scrubbed with a vegetable brush to remove any grit. The ends must be trimmed and discarded, but there is no need to remove the peel. In fact, in my opinion, the peel has the deepest flavor. Unless they are very dirty, winter squash need only a quick wipe and the removal of their seeds and fibers before cooking. Cut them in half, then scrape out the seeds and fibers with a very sharp spoon (a grapefruit spoon with a serrated edge is ideal). Winter squash that are grated or chopped before cooking must be peeled. If you are cooking winter squash whole or simply cut in half, there is no need to remove the peel ahead of time; the skin comes off more easily after baking.

Storing: Summer squash keep best in the refrigerator, where they will last for a week or more in a plastic bag or in the crisper. Winter squash have a shelf life only slightly less enviable than a Twinkie's: They will last up to six months if kept dry and at a cool temperature—but not in the refrigerator. (I like to keep a selection in a copper bowl on my kitchen counter for both dinner and decoration.) Fresh pumpkin is the one exception: It should be cooked, mashed, drained well and frozen for prolonged storage.

Cooking: Summer squash can be transformed by just about any method, from grilling to high-heat roasting. Winter squash are best—in flavor and texture—when baked slowly in a 350 degree F. oven. Butternut and similar winter squash can be sliced in half lengthwise, scraped clean of seeds, then laid cut side down in a glass baking dish filled with water to a depth of approximately 1/2 inch. Pumpkin should be cut in half, seeded and wrapped in foil and baked on a baking sheet at 350 degrees F. Spaghetti squash, by contrast, should be left whole, pricked all over with a knife and baked dry and whole for an hour or so. When it is done, it can be sliced in half lengthwise, the seeds scooped out and the strands of flesh separated with a fork.

I don't recommend steaming any type of winter squash because it saps flavor and adds moisture; you might as well buy it frozen. And my one experience with squash in a microwave, on a fishing boat in Alaska, convinced me that it may be quick but it's also devastating. Good butternut emerged a waterlogged mess, and even steaming would have been preferable.

A special note on pumpkin: Fresh pumpkin flesh is much paler and wetter than canned pumpkin, which has had up to 65 percent of its moisture extracted in processing. Fresh and canned pumpkin are interchangeable in recipes, but be aware that baked items and ravioli will be more watery if made with fresh pumpkin.

Toasting Squash Seeds: To toast pumpkin seeds, or any other type of winter squash seeds, scoop out the slimy seeds, rinse well under water and remove all the membrane material. Blot dry with paper toweling. Spread a single layer of the seeds on a heavy, rimmed baking sheet and toast in a preheated 350 degree F. oven until the seeds are crisp and toasted, approximately 20 to 30 minutes. Let cool slightly, then crack open the outer shells and remove seeds. Cool and store in an airtight container.

Growing Your Own Squash: Squash is one of those garden gifts that keeps on giving. Not only is it easy to cultivate, even in a backyard, but as any zucchini planter knows, it is also notoriously prolific. A gardener can snip blossoms off vines all summer long and still harvest a bumper crop of squash in the fall.

Squash was growing in North America when the first Europeans arrived, and growers say it will still thrive like a native plant (if not a weed) if you let it. Native Americans traditionally planted it alongside corn and beans, using fish as fertilizer. The bean vines grew up the corn stalks and the squash vines ran along the ground in between.

Today, the best way to grow squash is to start the seeds, both summer and winter varieties, in individual peat moss trays about a month before your anticipated planting date. The seedlings can then be planted about five feet apart, to give the vines room to spread, and in mounds, so the soil will drain. (Some seed companies recommend planting basil between squash plants to economize on space; the herb can be harvested before the vines crowd in.)

Summer squash should be planted in early June and will be ready for first harvest by mid-July; winter squash can be planted earlier, in May, and will be ready by September, depending on the weather. Sunny days and cool nights are crucial for ripening winter squash. Both types should be watered regularly.

All squash produce blossoms, but if you want to pick squash blossoms to eat, it is better to harvest the males, which do not bear fruit. (However, be sure to keep a few males intact to insure future pollination.) Males can be distinguished by their narrow stems; the females can be detected by a bulge at the bottom of the blossom which is actually the beginnings of a baby squash.

Summer squash should be picked when they are no more than three to four inches long, and in fact they should be picked daily, since they can grow astronomically in a short period of time. The flavor is far more subtle in lilliputian zucchini. When they grow to the size of a pitcher's forearm, they tend to be stringy and less satisfying. Winter squash, however, should always be allowed to ripen fully. They're ready to pick when the skin is completely hard.

To avoid squash overkill, seed companies advise planting several varieties. You won't get sick of zucchini so fast if you grow a few sunburst or yellow crookneck as well.

Summer Squash:

Chayote: Available year-round in some supermarkets and in stores catering to Mexican or Caribbean customers. Pale green and pear shaped, almost like an albino avocado, this sweet and mild squash is popular in Mexican and Cajun cooking. Also known as mirliton, christophine, *chocho* and vegetable pear. Great simply steamed. Usually has to be peeled, but the seed is edible—and quite good.

Cymling: Available in summertime at farmers' markets and supermarkets. Pale to medium green, small and flat, this bowl-shaped squash has a zucchini-like flavor and is great fried. Also known as pattypan. Sometimes confused with scallopini, a dark green variety, which is actually the cross between a cymling and a zucchini.

Lita: Available in late summer, primarily at farmers' markets. This baseball bat–shaped pale green squash is an Italian cousin of zucchini with a discernibly sweeter flavor. Best sautéed or steamed but welcome anywhere zucchini would be.

Roly Poly: Available in late summer at farmers' markets. Perfectly round, deep green with yellow flesh, this squash looks like a miniature watermelon but is essentially a globular zucchini. The seeds are more noticeable than those of zucchini, but the flesh is tender and nutty-tasting. Great simply halved and baked, or it can be hollowed out and stuffed with rice and chorizo like a sweet dumpling, acorn or delicata.

Squash Blossom: Available in summertime at farmers' markets. This delicate, bright-yellow flower is best eaten the same day it is picked. The flavor is very mild but rewarding, especially when fried. Also great in soups or stuffed with cheese and baked.

Sunburst: Available in late summer at farmers' markets and, in miniature form, year-round in specialty markets. Bright yellow, shaped somewhat like a top, this nutty-tasting squash is great for stuffing. In baby form, it's best steamed.

Chayote

Cymling

Roly Poly

Crookneck

Zucchini

Straightneck

Yellow Squash: Available in summertime at farmers' markets and supermarkets in two varieties—crookneck and straightneck; the names reflect the shape of the squash. Both have a mild, zucchini-like flavor and can go anywhere zucchini would. Crookneck squash is very popular in southern kitchens and actually tastes best when cooked until mushy and combined into other dishes.

Zucchini: Available year-round at farmers' markets and supermarkets. Easily the most ubiquitous (if not inescapable) squash, known to the French and British as courgette, zucchini is the vegetable for all seasons. This long cylindrical-shaped squash typically has a light to dark green skin but is also bred to produce a golden yellow variety. Also available in miniature form. It's good in soups, salads and pastas, grilled and even in cookies.

Winter Squash:

Acorn: Available year-round in most supermarkets. This acorn-shaped squash is generally blue-green or orange on the outside with golden orange flesh, and has a mild flavor nicely accentuated by cinnamon and honey or brown sugar. A pale yellow variety called Table Queen has a somewhat dry flesh with a sweet flavor. Great for stuffing with apples or cheese.

Australian Blue: Available primarily from August through March in upscale markets. Essentially pumpkin from down under, this large squash is grayish blue-green with orange flesh, and has a mild pumpkin flavor. Can be used in place of pumpkin or buttercup. Ideal for soups, pies and other baked goods.

Buttercup: Available year-round in some supermarkets and most health-food stores and in late fall at farmers' markets. This round green squash with a tiny topknot has a dense flesh and a wonderfully full-bodied flavor that is much sweeter than other winter varieties. Perfect as a puréed side dish, as well as in soups, stuffed and in baked items.

Butternut: Available year-round in most supermarkets. Shaped something like a bowling pin, this tan squash has a superb creamy flavor and yields much more meat than other squash. Ideal for soups and baking and in ravioli or risotto. Can also be peeled, grated and sautéed.

Calabaza: Available year-round primarily in markets catering to Hispanic or West Indian clienteles. Huge round or pear-shaped squash with mottled skin and deep orange flesh. Most often sold in precut chunks. Ideal grated in fritters, in soups and as a purée.

Cucuzza: Available June through October in specialty markets and some groceries. An Italian import also known as snake squash, this pale green variety has a mild flavor and grows up to three feet long. Must be peeled and usually should be seeded. Cucuzza can be used two ways: If the seeds are soft, it can be steamed or used in most zucchini recipes; if the seeds are hard, use like a winter squash either baked, stuffed or stewed in chunks.

Delicata: Available late July through October in some supermarkets, farmers' markets and health-food stores. This slender, oval squash has a green-striped shell and a sweet yellow flesh

Acorn

Australian Blue

Buttercup

Butternut

Calabaza

Delicata

Cucuzza

Golden Nugget

Hokkaido

Hubbard

Kabocha

Pumpkin

Jack Be Little

Spaghetti

Sweet Dumpling

Turban

Winter Melon

that distinctly evokes the flavor of corn. Best savored simply baked and buttered but can also be stuffed.

Golden Nugget: Available year-round in most supermarkets. Small orange pumpkin-shaped squash with yellow flesh, this has a slightly sweet flavor. Great for stuffing with rice, like a sweet dumpling, or simply baked.

Hokkaido: Available early fall through winter in farmers' markets, health-food stores and some supermarkets. A Japanese variety with a deep orange shell, Hokkaido has an assertive flavor and dark orange flesh. Especially good in soups, tarts and risotto.

Hubbard: Available late fall through winter in most supermarkets. This oversized squash, has a blue-gray or orange shell and pale yellow flesh. The flavor is extremely mellow; the texture is mealier than other squash. Good for soups, baked goods and as a purée.

Kabocha: Available early fall into winter in upscale supermarkets or health-food stores. Green, blue-gray or deep orange on the outside, this has a golden flesh and rich flavor. Use in any dish in which buttercup would work, such as soups, purées and desserts.

Pumpkin: Available October through November at farmers' markets and some supermarkets. Three varieties are often marketed: Pie pumpkins, also called sugar pumpkins, are round, deep orange and are edible cousins to jack-o'-lanterns; miniatures, or Jack Be Littles, are dwarf versions of pie pumpkins and are perfect for stuffing; cheese, or Dickinson pumpkins, are a pale orange variety that can reach 600 pounds or more and are grown for the canning market. Great in soups, pies, muffins, breads, even risotto and pasta. Canned pumpkin is perfectly acceptable for most dishes.

Spaghetti: Available year-round at most supermarkets. This variety, the size of a football, with a pale yellow shell, is like a cross between a winter and a summer squash, with flesh that separates into pasta-like strands after cooking. Serve it like spaghetti or mix it with cheese and bake or fry it.

Sweet Dumpling: Sold year-round in upscale supermarkets and health-food stores, and in the fall at farmers' markets. A small squash with a green-and-white-striped shell and sublimely sweet orange-yellow flesh. Also ideal for stuffing or for savoring by itself, baked in halves.

Turban: Sold year-round in many supermarkets and health-food stores, and in the fall at farmers' markets. Just as its name implies, this orange-and-green-skinned squash has a topknot. The shape is dramatic for serving soups; the richly flavored flesh can be used anywhere pumpkin or butternut would be, in baked goods, in risotto or purées.

Winter Melon and Other Asian Squash: Available in fall and winter in Asian markets. Winter melon is a huge pale green squash. It is usually sold cut into manageable chunks and is mostly used in soups. Other Asian squash include the bottle gourd, which is similar to a summer squash; the fuzzy melon, a close relative; and the snake gourd, a cylindrical squash with a zucchini-like flavor. These, too, are best used in soups.

OPENERS

To me, the first course is the most inviting. I could live on appetizers, on what normal eaters consider finger foods. With that bent, and with the versatility of squash, it has been easy over the years to come up with myriad ways to start dinners and liven up parties by putting squash on the menu.

In summer or winter, there's no better foundation for a satisfying soup, whether simmered with corn and smoked trout in a chowder or puréed with aromatic vegetables and garnished with toasted pecans. Salads are just as suited to squash, particularly in summer, when roasted zucchini can be teamed with other garden bounty.

For parties, sliced summer squash become the base for crostini topped with goat cheese and tapenade. Miniature squash make great vehicles for creamed blue cheese with a hint of chives.

Because squash is so inherently healthful, I also deep-fry it with impunity. Grated calabaza retains its texture and flavor in a fritter batter with salt cod. But the quintessential squash appetizer is a blossom, battered lightly and cooked to a golden crisp in crackling oil. After a few of those, I may not even need a main course, let alone dessert.

Appetizers are normally dainty, meant to be eaten in small bites. But when they're made from squash, they're serious food.

Chilled Curried Sunburst Soup with Tomato and Avocado

This velvety soup is just as soothing when served hot, but the coolness accentuates the subtle flavors. Zucchini, lita or yellow squash can be substituted for the sunburst.

2 tablespoons unsalted butter
1 large Spanish onion, finely diced
2 cloves garlic, crushed
1 fresh jalapeño chili, seeded and chopped
6 medium sunburst squash, trimmed, halved and
    thickly sliced
1 tablespoon curry powder
1/2 teaspoon cumin seeds
3 cups turkey or chicken stock, preferably homemade
1 cup light cream or half-and-half
Salt and freshly ground white pepper, to taste
Diced ripe avocado and tomato, for garnish

In a soup pot over medium heat, melt the butter. Add the onion, garlic and jalapeño and sauté 10 minutes, or until soft. Add the squash, curry powder, cumin seeds and stock, stir well and bring to a boil. Lower the heat to medium-low and simmer uncovered approximately 30 minutes, or until the squash is very soft.

Let cool slightly, then working in batches, transfer to a blender and purée until smooth. Return the purée to the pot and stir in the cream. Heat gently, but do not allow to boil. Transfer to a bowl, cover and refrigerate to chill.

Season with salt and pepper just before serving. Ladle into individual bowls and garnish with diced avocado and tomato. *Serves 6*

Chilled Curried Sunburst Soup with Tomato and Avocado

Creamy Butternut Bisque

Because the squash in this soup is cooked on the stove-top, it proves great taste does not have to eat up a whole evening. This can also be made with pumpkin, buttercup or Hubbard squash.

Handful of pecans, for garnish
4 tablespoons (1/2 stick) unsalted butter
2 medium leeks, white parts only, carefully washed
    and coarsely chopped
1 cup coarsely chopped celery stalks and leaves
2 medium carrots, coarsely chopped
1-inch piece fresh ginger, peeled and roughly chopped
1 fresh jalapeño chili, seeded and roughly chopped
1 medium butternut squash (approximately 2
    pounds), peeled, seeded and cut into rough cubes
4 cups chicken stock, preferably homemade, or water
1/2 cup light cream or half-and-half
Salt and freshly ground white pepper, to taste

Preheat the oven to 375 degrees F.

Spread the pecans on a baking sheet and toast in the oven for 5 minutes. Let cool slightly, then chop finely. Set aside.

In a soup pot over medium heat, melt the butter. Add the leeks, celery, carrots, ginger and jalapeño and sauté approximately 15 minutes, or until wilted. Stir in the squash and stock and bring to a boil. Reduce the heat to medium-low, cover and simmer 20 to 30 minutes, or until all the vegetables are very soft.

Let cool slightly, then working in batches, purée in a blender until smooth. Return the purée to the pot and stir in the cream. Heat through, but do not allow to boil. Season with salt and pepper. Ladle into individual bowls, sprinkle with pecans and serve. *Serves 6 to 8*

Summer Squash and Corn Chowder with Smoked Trout

Since I like chowder but am not crazy for gritty clams, I put together this height-of-summer combination of crunchy zucchini and yellow summer squash with sweet corn and chunks of smoked trout. The ingredients may be nontraditional, but the flavor is classic.

6 slices smoked bacon
1 small yellow onion, cut into 1/2-inch dice
1 small red bell pepper, cored, seeded and cut into
    1/2-inch dice
3 ears of corn, shucked
1 cup water
2 cups milk
1 teaspoon dried thyme, crumbled
Dash of hot-pepper sauce

4 small, slender green zucchini, trimmed and cut into
    1/2-inch dice
2 small yellow summer squash, trimmed and cut into
    1/2-inch dice
3/4 cup heavy cream
2 small smoked trout fillets, skinned, any errant
    bones removed, and flaked
1/2 teaspoon Worcestershire sauce
Salt and freshly ground black pepper, to taste

In a soup pot over low heat, fry the bacon 5 to 10 minutes, or until crisp. Remove the bacon from the pan and drain on paper towels.

Pour off all but 1 tablespoon of the bacon drippings. Place the soup pot over medium-low heat. Add the onion and red pepper and sauté 10 minutes, or until tender. Using a sharp knife, cut off the corn kernels from the ears. Add the kernels to the pot along with the water, milk, thyme and hot-pepper sauce. Simmer, uncovered, over medium-low heat approximately 10 minutes, or until the corn is tender.

Stir in the zucchini and yellow squash and continue simmering approximately 20 minutes longer, or until all the vegetables are soft.

Stir in the cream, trout and Worcestershire sauce and heat through, but do not allow to boil. Season with salt and pepper. Ladle into individual bowls and crumble the bacon over the top and serve. *Serves 4*

Caramelized Pumpkin and Onion Soup with Smoked Turkey

*This variation on classic pumpkin soup takes a little more time than the butternut bisque
but has more flavor nuances. It is rich but cream-free and looks spectacular served in a hollowed-out
winter squash shell. A turban squash can be used in place of the pumpkin.*

1 large pie pumpkin (approximately 3 pounds)
1 teaspoon plus 3 tablespoons extra virgin olive oil
2 large yellow onions, very thinly sliced
2 carrots, cut into fine julienne strips
1 tablespoon granulated sugar
1/2 teaspoon ground sage

4 cups turkey or chicken stock, preferably homemade
Salt and freshly ground white pepper, to taste
1/4 pound high-quality smoked turkey, cut into
    strips 2 inches long by 1/4 inch wide
1/2 cup grated Gruyère cheese

Preheat the oven to 375 degrees F.

Cut the pumpkin in half and scrape out the seeds. Brush the cut surfaces lightly with the 1 teaspoon of oil, then lay cut side down in a glass baking dish. Bake in the oven approximately 1 hour, or until the flesh is very soft and the cut sides are slightly caramelized. Let cool, then scrape out the pulp.

While the squash bakes, warm the 3 tablespoons of oil in a large soup pot over medium heat. Add the onions and carrots and cook, stirring often, approximately 15 minutes, or until soft. Sprinkle with the sugar and continue cooking, stirring often, approximately 10 to 15 minutes longer, or until the vegetables are browned and caramelized.

Stir in the sage, stock and pumpkin pulp, and bring to a boil. Let cool slightly, then working in batches, transfer to a blender and purée until smooth. Return the purée to the soup pot and reheat gently, but do not allow to boil.

Season with salt and pepper. Ladle into individual bowls—or into a hollowed-out winter squash shell—and garnish with the turkey and Gruyère. *Serves 4 to 6*

Deep-Fried Squash Blossoms

*Zucchini blossoms are one of the great harbingers of summer. There is simply no other time to
indulge in the ephemerality of these fragile blossoms. If you aren't lucky enough to have a garden or a good
farmers' market, try the same fritter treatment on thin rounds of zucchini, lita or cymlings.*

*1/2 cup milk
1 extra-large egg, separated
1/2 cup all-purpose flour
1/4 teaspoon salt
1/4 teaspoon ground mace
2 to 3 dashes cayenne pepper*

*1 tablespoon corn oil, plus at least 4 cups corn oil,
    for deep-frying
1 clove garlic, minced
Approximately 3 dozen zucchini blossoms
Salsa (store-bought or homemade) or hot-pepper sauce*

In a medium bowl, beat together the milk and egg yolk. Whisk in the flour, salt, mace, cayenne, 1 tablespoon of the corn oil and garlic. Beat well.

In a separate small bowl, beat the egg white until stiff but not dry. Gently fold into the batter and set aside.

In a deep, heavy skillet, pour in oil to a depth of 3 to 4 inches. Heat the oil until the temperature registers 360 degrees F. on a frying thermometer. One at a time, dip the squash blossoms into the batter, coating completely, then slip them into the hot oil. Do not crowd the pan. Fry, turning once, 2 to 5 minutes on a side, or until crisp and brown. Using a slotted spoon, remove to paper towels to drain briefly. Serve hot with plenty of salsa for dipping. *Makes approximately 3 dozen blossoms*

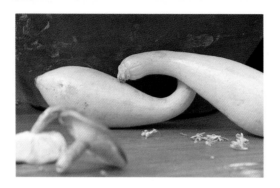

Baby Squash Stuffed with Chives and Maytag Blue

In this appetizer, squash and blue cheese are a marriage made in Iowa. The pungent
Midwestern cheese plays perfectly against the nutty-tasting baby summer squash. Other blue cheeses,
such as Gorgonzola or French or Danish blues, can be substituted for Maytag.

8 baby yellow crookneck squash, or green or golden
  zucchini, each 3 to 4 inches long
One 3-ounce package cream cheese, softened
3 ounces Maytag blue cheese, softened

1 tablespoon snipped fresh chives, plus whole chives,
  for garnish
1 to 2 dashes hot-pepper sauce

Fill a saucepan three-fourths full of water and bring to a boil. Drop in the squash and boil for 1 minute. Drain, then immediately plunge the squash into ice water to stop the cooking. Drain again and dry completely with paper towels. Trim the ends, then cut in half lengthwise. Using the small end of a melon baller, or a similar tool, carefully hollow out the center of each squash half, leaving a shell 1/8 inch thick. Drain upside down on paper towels until all moisture is drained off, approximately 10 minutes. (The small amount of squash removed from the center can be discarded or reserved for another use.)

In a bowl and using a wooden spoon, stir together the cream cheese and blue cheese until smooth. Stir in the snipped chives and hot-pepper sauce, to taste.

Using a miniature spatula, spread the cheese mixture into the hollowed-out squash. Arrange the squash on a plate and garnish with whole chives. Cover and refrigerate until chilled before serving. *Serves 4 to 6*

Bajan Calabaza and Cod Fritters

*Every beachside vendor in Barbados offers some variation on these succulent fritters with
shreds of orange squash set off against white salt cod. They make a superb accompaniment for cocktails,
especially when teamed with incendiary West Indian–style hot-pepper sauce.*

1/2 pound boneless salt cod
1 cup all-purpose flour
1 teaspoon baking powder
1 large egg, lightly beaten
1/2 cup milk
2 tablespoons corn oil, plus at least 4 cups corn oil,
    for deep-frying

1 cup loosely packed, coarsely grated calabaza,
    pumpkin or turban squash
4 green onions, green parts only, finely chopped
1 clove garlic, minced
1 fresh Scotch Bonnet or serrano chili, seeded and
    finely chopped

Place the cod in a large glass bowl and add cold water to cover. Soak for at least 8 hours in the refrigerator, changing the water at least three times. Drain well, then place in a large, shallow saucepan and cover with boiling water. Cover and steep 20 minutes, or until soft. Drain and cool, then shred the cod into fairly fine flakes.

In a medium bowl, stir together the flour and baking powder. Make a well in the center and add the egg, milk and the 2 tablespoons corn oil. Using a wooden spoon, gradually stir the flour into the wet ingredients, mixing until blended. Stir in the cod, squash, onions, garlic and chili until thoroughly combined.

In a Dutch oven or deep, heavy skillet, pour in oil to a depth of 3 inches. Heat to 360 degrees F. on a frying thermometer. Working in batches, drop the batter, by teaspoonfuls, into the hot oil and fry 4 to 5 minutes, or until crisp and golden. Using a slotted spoon, remove to paper towels to drain briefly, then serve hot. *Makes approximately 4 dozen fritters*

Savory Zucchini and Cheese Madeleines

*A French madeleine mold converts a quiche-like filling into a savory
hors d'oeuvre with a crunchy crust and moist center. If you don't have a madeleine mold,
use miniature muffin tins and bake for approximately 15 minutes.*

2 large eggs
2 tablespoons heavy cream
*2 teaspoons Creole or Pommery (coarse-grain)
    mustard*
2 tablespoons unsalted butter, melted
3 cloves garlic, minced
1 1/2 teaspoons dried basil, crumbled
1 teaspoon salt
1/4 teaspoon freshly ground black pepper

Dash of cayenne pepper
1 cup all-purpose flour
1/2 teaspoon baking powder
1/4 cup coarse-grind yellow cornmeal
*2 cups firmly packed, coarsely grated green zucchini
    (approximately 3 medium squash)*
1 small yellow onion, finely diced
1 small red bell pepper, cored, seeded and finely diced
1 cup grated Gruyère or Jarlsberg cheese

Preheat the oven to 425 degrees F. Oil 3 madeleine molds and set aside. (If you only have 1 mold, work in batches, letting the mold cool before refilling.)

In a large bowl, combine the eggs, cream, mustard, melted butter, garlic, basil, salt and black and cayenne peppers and mix well. Stir in the flour, baking powder and cornmeal and mix well. Add the zucchini, onion, bell pepper and cheese and mix thoroughly. Spoon into the prepared madeleine molds.

Bake in the oven 20 minutes, or until puffed and golden brown (the centers will still be moist). Turn out of the molds and serve warm, or let cool on wire racks to room temperature. *Makes approximately 3 dozen madeleines*

Lita Crostini with Goat Cheese and Tomato Tapenade

There is no need to invest in crackers in summertime when squash are so cheap and plentiful.
Thin rounds of crunchy lita make an ideal platform for a pungent paste accented by goat cheese.
Zucchini, yellow squash or even sunburst can be substituted for the sweet lita.

1/4 cup sun-dried tomatoes (not oil-packed)
1/4 cup pitted Kalamata olives
1/4 cup firmly packed fresh basil or parsley leaves
1/2 teaspoon Worcestershire sauce
1 teaspoon anchovy paste
2 tablespoons extra virgin olive oil

1 clove garlic, minced
Salt and freshly ground black pepper, to taste
4 to 6 medium lita squash
One 3 1/2-ounce round soft, mild goat cheese
2 tablespoons light cream
Chopped fresh basil or parsley, for garnish

Place the sun-dried tomatoes in a small bowl and cover with boiling water. Let stand approximately 15 minutes, or until soft. Drain well and transfer to a blender or to a food processor fitted with the metal blade. Add the olives, basil leaves, Worcestershire sauce, anchovy paste and oil. Purée to form a fairly smooth paste, scraping down the sides of the container frequently. Stir in the garlic and season with salt and pepper. Set aside.

Trim the ends of the squash and cut into rounds approximately 1/4 inch thick. Place the cheese in a small bowl. Add the cream and stir to thin the cheese to a smooth, spreadable consistency. Spread a little of the cheese over each squash round. Top with a teaspoon or so of the tomato-olive mixture. Garnish with the chopped basil or parsley and serve. *Makes approximately 4 dozen crostini*

Roasted Green and Golden Zucchini Salad
with Peppers and Mozzarella

*Roasting is a world-class way to transform everyday squash into a
tantalizing vegetable, especially in a salad. This can be a great fresh-from-the-garden
main course for a summer lunch as well as an appealing antipasto.*

2 medium red bell peppers
4 medium, slender green zucchini
4 medium, slender golden zucchini
1 large red onion, halved and thickly sliced
3 tablespoons extra virgin olive oil
1 teaspoon coarse sea salt
1/8 teaspoon freshly ground black pepper

1 small head red-leaf lettuce, leaves separated,
　washed and dried
1/4 pound mozzarella cheese, preferably fresh
1/4 cup halved, pitted Niçoise olives
2 tablespoons balsamic vinegar
8 fresh basil leaves

Preheat the broiler.

Arrange the bell peppers on a heavy baking sheet and place under the broiler. Roast, turning to expose all surfaces, until they are evenly charred and blistered on all sides. (The roasting time will vary depending on the thickness of the peppers and the intensity of the broiler; figure on 15 to 25 minutes.) Let cool, then slip off the skins, core and remove the seeds. Cut into strips approximately 3/4 inch wide and 2 to 4 inches long; set aside. Adjust the oven setting to 450 degrees F.

Trim the ends of the zucchini. Cut in half crosswise and then into quarters lengthwise. Place in a large mixing bowl. Add the onion slices and 1 tablespoon of the oil. Add the salt

and pepper and toss with your hands until the vegetables are lightly and evenly coated with the oil. Spread them out in a single layer on a large, heavy baking sheet. Place in the oven and roast, turning often, 15 to 20 minutes, or until all the vegetables are soft. Let cool slightly.

Line a shallow serving dish or a platter with the lettuce. Cut the mozzarella into strips approximately the same size as the pepper strips. Arrange the zucchini and onion slices, roasted pepper strips and mozzarella atop the lettuce. Scatter the olives over the vegetables. Drizzle with the remaining 2 tablespoons oil, and then with the vinegar. Stack the basil leaves and cut crosswise into very thin strips. Sprinkle them over the salad and serve. *Serves 4*

Kabocha Corn Muffins with Sweet Onions

*Moist and tender, these easy-to-make muffins are great with soups, for brunch or as a snack,
served hot with butter. To serve them as an hors d'oeuvre with drinks, add a cupful of
finely diced country ham to the batter and bake in miniature muffin tins for 20 minutes, or until set.
Mashed pumpkin, butternut, buttercup, acorn or Hubbard squash can replace the kabocha.*

1 1/2 cups coarse-grind yellow cornmeal
1/2 cup all-purpose flour
1 1/2 teaspoons baking powder
1 teaspoon baking soda
1 teaspoon salt
1 tablespoon chopped fresh sage, or 1/2 teaspoon
    dried sage, crumbled

1 teaspoon chopped fresh rosemary, or 1/4 teaspoon
    dried rosemary, crumbled
1 small Vidalia or other sweet onion, finely diced
2 large eggs
3/4 cup sour cream or plain yogurt
4 tablespoons (1/2 stick) unsalted butter, melted
1 cup mashed, cooked kabocha squash

Preheat the oven to 425 degrees F. Butter a 12-cup muffin tin and set aside.

In a large bowl, stir together the cornmeal, flour, baking powder, baking soda, salt, sage and rosemary. Stir in the onion and set aside.

In a second bowl, beat the eggs lightly, then stir in the sour cream and melted butter. Whisk in the squash until smooth.

Add the squash mixture to the cornmeal mixture and stir to combine fully. Spoon into the prepared muffin cups, until each cup is two-thirds full. Bake in the oven 25 to 30 minutes, or until golden brown and a toothpick inserted in the center comes out clean. Let cool a few minutes, then turn out of the tin and serve hot. *Makes 1 dozen muffins*

Pesto and Zucchini Triangles

*Zucchini and basil are natural partners, but this little hors d'oeuvre pushes the concept further
with the addition of garlic and Fontina cheese. If you don't want to mess with phyllo dough, try
wonton wrappers (they are sold in some produce sections) but cut way back on the butter.*

1 cup firmly packed fresh basil leaves
1/2 cup walnuts
1/4 cup extra-virgin olive oil
3 cloves garlic, minced
Dash of hot-pepper sauce

1 cup grated Fontina cheese
1 1/2 cups coarsely grated green or golden zucchini
Salt and freshly ground black pepper, to taste
8 sheets phyllo dough
8 tablespoons (1 stick) unsalted butter, melted

Preheat the oven to 375 degrees F. Butter 1 or 2 baking sheets.

In a blender, combine the basil, walnuts and olive oil and purée until fairly smooth. Transfer to a bowl and mix in the garlic and hot-pepper sauce. Add the Fontina and the zucchini and mix well, then season with salt and pepper.

Butter a work surface and lay out 1 phyllo sheet lengthwise; keep the others loosely covered with a barely damp towel. Brush the phyllo with melted butter and fold in half horizontally. Using a sharp knife, cut phyllo vertically into 5 strips, each approximately 2 1/2 inches wide. Place 1 1/2 teaspoons of the zucchini mixture near the bottom right corner of each strip and fold upward and over on the diagonal to form a triangular shape. Bring the bottom tip of the triangle up to align with the straight edge. Fold again on the diagonal and continue in this manner, as if folding a flag, until you have an enclosed triangular pastry. Place on the prepared baking sheet. Repeat with the remaining phyllo and filling.

Bake in the oven 20 to 25 minutes, or until golden brown. Serve warm. *Makes 40 triangles*

ACCOMPANIMENTS

Squash is my idea of a kitchen chameleon. Any one of the winter varieties can be transformed through baking, stuffing or puréeing, while summer squash are suited for the quick-change artistry of roasting, grilling or sautéing. This versatility makes squash the quintessential side dish. It can take on the flavor of any cuisine, whether all-American or completely ethnic, and there isn't a main course it doesn't complement.

Most squash in either the winter or summer category can be treated interchangeably, but each one has its own distinguishing characteristics. Chayote, for instance, is so mellow it benefits from simple steaming and glazing with a little cream and herb. Zucchini is ideal fried, but it doesn't have to be fried in a skillet; it gets just as crunchy in the oven coated in a Parmesan batter. And delicata is best baked slowly to bring out the subtle corn flavor.

If you've never tried a particular squash, though, the first way to savor it is always the simplest. Steam or sauté a summer squash; slowly bake a winter one. They all have individual flavors, and once you get to know them you can refine the preparations.

The one treatment not included in this section is a classic winter squash baked with cinnamon and brown sugar. I'd rather play off the sweetness, not accentuate it. Any squash can be sugary. Why not put it to the test?

Oven-Fried Zucchini in a Crunchy Parmesan Crust

Zucchini is traditionally fried in a hot skillet in lots of oil, but cooking it in a crust in the oven leaves it less greasy and far more savory. These can be served hot as a side dish or at room temperature as an hors d'oeuvre. Sticks or rounds of other summer squash, such as cymling, lita or crookneck, can be prepared the same way.

1 tablespoon extra virgin olive oil
1/4 cup fine dried bread crumbs
1/3 cup grated imported Parmesan cheese
1/2 teaspoon dried rosemary, crumbled
2 to 3 dashes cayenne pepper

1/2 teaspoon salt
1/4 teaspoon freshly ground black pepper
1 large egg
4 small green or golden zucchini

Preheat the oven to 400 degrees F. Lightly grease a heavy baking sheet with the oil and set aside.

In a shallow dish, combine the bread crumbs, Parmesan, rosemary, cayenne, salt and pepper and mix well. In a second shallow dish, lightly beat the egg.

Trim the ends of the squash. Cut each squash in half lengthwise. Lay the halves flat and cut in half lengthwise again. Then cut the strips in half crosswise. Dredge each piece first in the egg and then in the Parmesan mixture, coating evenly. Arrange well spaced in a single layer on the prepared baking sheet. Bake in the oven for 5 to 7 minutes, then turn the squash over and bake 5 to 7 minutes longer, or until crisp and lightly browned. Serve hot or at room temperature. *Serves 4*

Southern Crookneck Soufflé

*Not as risky, or as high-rising, as a real soufflé, this Georgia classic makes a rich combination
of ingredients seem light and airy but wholly satisfying. The finished dish is something like spoon bread.
In winter, it can be made with cooked butternut or buttercup squash in place of the crookneck.*

*6 medium yellow crookneck squash, trimmed and
   thinly sliced*
1 cup heavy cream
2 large eggs, beaten

1/2 teaspoon salt
1/4 teaspoon freshly ground white pepper
1/8 teaspoon ground mace
1/2 cup grated sharp Cheddar cheese

Preheat the oven to 350 degrees F. Butter a
1 1/2-quart soufflé dish or casserole.

Bring water to a boil in a steamer pan,
place the squash slices on the steamer rack,
cover and steam approximately 15 minutes, or
until very soft.

Drain well in a colander and transfer to a large
bowl. Using a potato masher, mash the squash
until fairly smooth. Stir in the cream and eggs
and season with the salt, pepper and mace. Pour
into the prepared dish and sprinkle the cheese
evenly over the top.

Bake in the oven approximately 25 minutes, or until set. Serve at once. *Serves 4 to 6*

Zucchini, Tomato and Basil Gratin

*A top-of-the-season side dish, this easy assemblage looks gorgeous and tastes
great on a summer menu that incorporates anything from grilled steaks to potato salad.
You can also try making it using other herbs, such as cilantro or rosemary.*

2 medium green zucchini, each approximately
    1 1/2 inches in diameter
6 ripe plum tomatoes
6 large fresh basil leaves

1 large clove garlic, halved
3 teaspoons extra virgin olive oil
Salt and freshly ground black pepper, to taste
1/4 cup grated imported Parmesan cheese

Preheat the oven to 350 degrees F.

Trim the ends of the zucchini. Cut into slices 1/4 inch thick. Cut the tomatoes into slices 1/4 inch thick as well. Stack the basil leaves and cut into strips approximately 1/4 inch wide. Set aside.

Rub the garlic clove over the bottom of a shallow 10-inch-square glass or ceramic baking dish. Grease with 1 teaspoon of the oil.

Arrange the zucchini and tomato slices in a single layer in the dish, overlapping them tightly. Season well with salt and pepper. Scatter the basil strips evenly over the top, then drizzle with the remaining 2 teaspoons oil. Finally, sprinkle the Parmesan over the top.

Bake in the oven for 20 minutes, or until the zucchini is tender. Serve hot or at room temperature. *Serves 4*

Grilled Summer Squash with Herbs

Grilled squash is one of the greatest additions to the backyard barbecue.
The charred flavor makes plain squash irresistible, not just as an accompaniment to other grilled
foods but also as an appetizer. Try it topped with a little goat cheese.

6 medium summer squash (such as sunburst, cymling
or zucchini)
2 tablespoons extra virgin olive oil
2 cloves garlic, minced

1 teaspoon kosher or coarse sea salt
1/4 teaspoon freshly ground black pepper
1 tablespoon coarsely chopped mixed fresh herbs
(such as basil, rosemary, dill and/or parsley)

Preheat a stove-top grill or prepare a fire in a charcoal grill.

Trim the ends of the squash. Cut lengthwise into slices approximately 1/2 inch thick (or, in the case of cymling or sunburst, cut horizontally). Place in a large, shallow bowl and add the oil and garlic. Season with the salt and pepper and toss until all the squash slices are well coated.

Working in batches, lay the squash slices on the grill. Cover and cook, turning once, approximately 3 to 5 minutes on each side, or until tender. Arrange the squash on a platter and sprinkle with the herbs. *Serves 4*

Shaker Yellow Velvet

*An old-time Shaker recipe, this sublime combination
of crookneck squash and corn makes an ideal partner
for simple grilled fish or roast chicken or pork.*

3 ears of corn, shucked
1/2 cup chicken stock, preferably homemade,
    or water
3 yellow crookneck squash, each approximately
    6 inches long
1 cup heavy cream
1 tablespoon snipped fresh chives
Salt and freshly ground white pepper, to taste

Bring water to a boil in a steamer pan. Place the
corn on the steamer rack, cover and steam 5
to 10 minutes, or until very tender. Let cool.
Using a sharp knife, cut off the kernels from
the cobs. Place the corn in a skillet with the
chicken stock or water and set aside.

Trim the ends of the squash. Cut into
quarters lengthwise, then cut crosswise into
slices 1/2 inch thick. Add to the skillet. Cover
and cook over medium heat approximately 10
minutes, or until the squash is tender.

Uncover and cook over medium-high
heat approximately 5 minutes longer, or until
all the liquid evaporates. Stir in the cream and
cook, stirring occasionally, approximately 5
minutes or so, or until the cream thickens.

Stir in the chives and season with salt and
pepper. Serve immediately. *Serves 4*

Steamed Chayote
in Cilantro Cream

*Chayote has the most subtle flavor of any
squash in any season, so it takes well to a touch of
cream with a hint of cilantro. This pale green
squash is best when cooked al dente, so take care
not to overcook it. Zucchini or cymlings could be
substituted, but the flavor is not as delicate.*

2 medium chayote squash
3/4 cup heavy cream
1 clove garlic, crushed
1/4 teaspoon pure chili powder
2 tablespoons coarsely chopped fresh cilantro
Salt and freshly ground white pepper, to taste

Cut each squash in half lengthwise. Using a
vegetable peeler, remove the skin. Cut the
squash into medium dice. Bring water to a
boil in a steamer pan, place the squash on the
steamer rack, cover and steam 10 to 15 minutes, or until just tender but still crisp.

Meanwhile, in a small saucepan, combine
the cream, garlic and chili powder. Place over
medium-high heat and cook approximately
10 minutes, or until thickened and reduced
by half. Remove and discard the garlic.

When the squash is cooked, transfer to a
bowl and pour the cream mixture over the
top. Add the cilantro and toss until evenly
coated. Season with salt and pepper and serve
immediately. *Serves 4*

Sunburst Rémoulade

Rémoulade is the traditional French and Creole treatment for celery root and shrimp, respectively, but it works even more harmoniously with julienned summer squash. This is essentially a citified coleslaw. Zucchini, lita or yellow squash can fill in for sunburst.

1/3 cup mayonnaise
1 tablespoon Creole or Pommery (coarse-grain) mustard
1 teaspoon anchovy paste (optional)
4 cornichons or tiny gherkins, finely diced
1 tablespoon capers, drained and coarsely chopped
1 tablespoon snipped fresh chives
2 teaspoons minced fresh parsley
1 to 2 dashes hot-pepper sauce
3 medium sunburst squash, cut into 1/4-inch matchsticks
1 medium, slender green zucchini, cut into 1/4-inch matchsticks
Salt and freshly ground black pepper, to taste
Lettuce leaves, for serving

In a medium bowl, combine the mayonnaise, mustard and the anchovy paste, if desired. Stir until well blended. Add the cornichons, capers, chives, parsley and hot-pepper sauce. Stir well until the ingredients are evenly distributed.

Place all the squash matchsticks in a large bowl and toss with the mayonnaise mixture until coated. Season with salt and pepper. Cover and refrigerate at least 1 hour to allow the flavors to blend.

To serve, line a shallow bowl with lettuce leaves and spoon the squash mixture into the center. *Serves 4*

Roasted Delicata with Red Chili and Lime Butter

Delicata squash are so satisfying they really need no gilding, but a hint of butter, chili powder and lime brings out extra taste in a squash with a gentle corn flavor. Sweet dumpling, golden nugget or roly poly squash can be substituted for the delicata.

2 delicata squash (approximately 3/4 pound each)
3 tablespoons unsalted butter, softened
1 tablespoon freshly squeezed lime juice
1 teaspoon pure chili powder
Salt and freshly ground black pepper, to taste

Preheat the oven to 350 degrees F.

Cut the squash in half lengthwise and scrape out the seeds. Place cut side down in a glass baking dish and add water to the dish to a depth of approximately 1/4 inch. Bake approximately 20 minutes, or until the squash is soft but not mushy.

In a small bowl, blend together the butter, lime juice and chili powder until thoroughly combined. Season with salt and pepper. Spoon the butter mixture into the delicata cavities and serve hot. *Serves 4*

Sunburst Squash Stuffed with Spinach and Gruyère

*These colorful stuffed squash are meant as a side dish, but they can also be served as
a brunch or lunch main dish. The spinach, cheese and shiitakes are good counterpoints to
the nutty-tasting sunburst. Try the same filling in large zucchini or yellow squash.*

4 medium sunburst squash, each 3 1/2 to 4 inches
    in diameter
2 tablespoons unsalted butter
1 tablespoon extra virgin olive oil
6 shiitake or button mushrooms, stemmed and
    finely chopped
2 shallots, finely chopped
1 clove garlic, minced

1 teaspoon soy sauce
1 large bunch fresh spinach leaves, carefully washed
    and finely chopped
1/8 teaspoon freshly grated nutmeg
Dash of cayenne pepper
Salt and freshly ground black pepper, to taste
1 large egg, lightly beaten
3/4 cup grated Gruyère or Swiss cheese

Preheat the oven to 375 degrees F.

Fill a large pot three-fourths full of water and bring to a rolling boil. Drop the squash into the boiling water and boil for 5 minutes. Drain well and let cool.

Cut a thin slice off the bottom of each squash, removing just enough so that it will stand upright. Then slice off the tops (approximately 1/2 inch thick) and hollow out the centers, leaving a thin shell. Set aside. (Reserve the nominal amount of squash removed from the centers for another use or simply discard.)

In a skillet over medium heat, melt the butter with the oil. Add the mushrooms, shallots and garlic and sauté approximately 10 minutes, or until soft. Stir in the soy sauce and add the spinach. Raise the heat to medium-high. Sauté for approximately 5 minutes, or until the spinach is tender and most of the liquid has evaporated. Transfer to a bowl and let cool slightly, then season with the nutmeg, cayenne, salt and black pepper. Stir in the egg and 1/2 cup of the Gruyère and mix well.

Mound the mixture into the hollowed-out squash. Arrange in a single layer in a baking dish just large enough to hold all the squash upright. Sprinkle the remaining 1/4 cup cheese over the tops. Pour hot water into the pan to a depth of 1/2 inch.

Bake in the oven for approximately 30 minutes, or until the filling is set and the squash is tender. Serve immediately. *Serves 4*

Fast and Spicy Shredded Buttercup Sauté

*The one drawback to winter squash—the long, slow cooking required—is short-circuited in
this quick and easy dish because grating brings the buttercup down to size. The spice of the hot peppers
makes it an appealing side dish. Butternut, kabocha or calabaza squash can be substituted here.*

1 large buttercup squash (approximately 3 pounds)
4 tablespoons (1/2 stick) unsalted butter
2 shallots, minced

1/4 teaspoon red pepper flakes
Salt and freshly ground black pepper, to taste

Peel the squash, cut in half and scrape out the
seeds. Grate the squash on the coarsest holes
of a four-sided grater. Set aside.

In a large, deep skillet, melt the butter over
medium heat. Stir in the squash, shallots and
pepper flakes. Cook, stirring constantly, 7 to 10
minutes, or until the squash is tender but not
mushy. Season with salt and pepper and serve
immediately. *Serves 4 to 6*

Thyme-Scented Turban Squash Purée

*This recipe sounds absurdly simple, and it is, but what emerges from the blender
is a world-class dish, velvety smooth and heady with thyme. It's ideal with everything from
roast turkey to fried chicken to grilled tuna. Any other winter squash will work.*

1 large turban squash (approximately 3 pounds)
4 tablespoons (1/2 stick) unsalted butter
1 tablespoon chopped fresh thyme, or 1 teaspoon
   dried thyme, crumbled

Dash of cayenne pepper
Salt and freshly ground white pepper, to taste

Preheat the oven to 350 degrees F.

Cut the squash in half and scrape out the seeds. Cover the cut side of each half with aluminum foil and place foil side up on a baking sheet. Bake in the oven for approximately 1 hour, or until very soft.

Uncover and scrape out the soft flesh into a blender or into a food processor fitted with the metal blade. Add the butter and thyme and purée until smooth, scraping down the sides of the container frequently. Season with the cayenne, salt and pepper.

Transfer to a serving dish and serve at once, or spread into a small baking dish and keep warm in a 250 degree F. oven until ready to serve. *Serves 4*

MAIN COURSES

One of my most memorable eating epiphanies was the first taste of pumpkin ravioli in an Italian restaurant. Until then, I had always thought of this classically American squash as a pie ingredient. The concept of making squash into a main course was completely foreign.

And traditionally it has been. While the earliest Americans treated squash almost like meat, successive generations have generally relegated it to the side of the plate, which is ironic, since the rest of the globe had been limping along without squash until the New World was discovered. Vegetarians might stuff an acorn squash with bulgar wheat and call it dinner, but otherwise only ethnic cooks have explored the full potential of squash. Once the Europeans found it they didn't let it go. In Italy, squash is tucked into tortellini and simmered in risotto. In France, zucchini is stewed with eggplant, peppers and tomatoes to make arguably the most satisfying vegetarian main course ever: ratatouille. And the Greeks skewer zucchini with lamb or mix it into meatballs.

The recipes in this section not only play off those ideas but also go back to squash's roots. Mexican enchiladas take on a new taste with a zucchini and cheese filling, while all-American pot pies are given new life with calabaza and turkey. When it comes to main courses these days, squash knows no boundaries.

Spaghetti Squash Primavera

*Spaghetti squash has a huge advantage over real pasta: It has negligible calories when compared
with authentic noodles. Add an array of other fresh vegetables and even the cream sauce tastes guilt-free.*

1 medium spaghetti squash (approximately 3 pounds)
2 cups heavy cream
1/2 cup chicken stock, preferably homemade
2 cloves garlic, crushed
1/4 teaspoon red pepper flakes
1 1/2 cups small broccoli florets
1 small golden zucchini, cut into 1/4-inch matchsticks
1 small green zucchini, cut into 1/4-inch matchsticks
1 1/2 cups fresh peas (or frozen peas, thawed)
2 tablespoons unsalted butter

*1 medium red bell pepper, cored, seeded and cut into
    julienne strips*
*1/2 pound shiitake mushrooms, stemmed and sliced
    1/4 inch thick*
1 tablespoon Creole or Pommery (coarse-grain) mustard
*1 cup grated imported Parmesan cheese, plus
    additional grated Parmesan, for garnish*
Salt and freshly ground black pepper, to taste
*2 tablespoons coarsely chopped mixed fresh herbs
    (such as basil, chives, dill and/or parsley)*

Preheat the oven to 350 degrees F.

Leaving the spaghetti squash whole, prick the skin all over with a knife. Place on an ungreased baking sheet without any water and bake in the oven for an hour.

While the spaghetti squash is cooking, in a heavy saucepan over medium heat, combine the cream, stock, garlic and pepper flakes. Bring to a boil, then reduce the heat to low and simmer, stirring occasionally, approximately 20 minutes, until thickened. Remove and discard the garlic. Remove the sauce from the heat and cover to keep warm.

While the sauce is thickening, fill a saucepan three-fourths full of water. Bring to a boil. Drop the broccoli into the water and boil for 1 to 2 minutes. Using a slotted spoon, remove the broccoli, drain well, and add to the sauce. Add the zucchini strips to the same boiling water and boil for 30 seconds. Then drain and add to the warm sauce. Add the peas to the boiling

water for 2 to 3 minutes. Drain and add to the sauce as well.

In a medium skillet over medium-high heat, melt 1 tablespoon of the butter. Add the bell pepper and sauté for approximately 5 minutes, or until tender. Transfer to the sauce. Add the remaining butter to the same pan over medium-high heat. Add the mushrooms and sauté approximately 10 minutes, or until browned and tender. Transfer to the sauce.

When the spaghetti squash is done, let cool slightly and slice in half lengthwise and scoop out the seeds. Use a fork to scrape out the strands into a large warmed pasta bowl. Separate the strands with the fork. Whisk the mustard and Parmesan into the sauce and reheat gently. Pour the sauce over the spaghetti squash and toss until the strands are evenly coated and the ingredients are mixed well. Season with salt and pepper. Sprinkle the herbs over the top and garnish with Parmesan. Serve immediately. *Serves 4*

Turkey and Calabaza Pot Pies with Parmesan Crust

*Pot pies get an update in this savory dish that evokes the flavor of Thanksgiving with
none of the heavy work. You can make these with leftovers, but smoked turkey gives them more class.
Pumpkin, butternut, turban or other winter squash can be substituted for the calabaza.*

Crust:

1 1/3 cups all-purpose flour
1/2 teaspoon salt
1/4 cup shortening, chilled
4 tablespoons (1/2 stick) unsalted butter, chilled
3 tablespoons ice water

Filling:

3 tablespoons unsalted butter
2 medium leeks, white parts only, carefully washed
    and thinly sliced
1 small red bell pepper, cored, seeded and cut into
    medium dice
2 celery stalks, trimmed and finely diced
1 1/2 cups peeled, 3/4-inch diced calabaza

2 cups turkey or chicken stock, preferably
    homemade, heated
1/3 cup all-purpose flour
1 tablespoon chopped fresh thyme, or 1 teaspoon
    dried thyme, crumbled
1 tablespoon chopped fresh sage, or 1/2 teaspoon
    dried sage, crumbled
2 to 3 dashes hot-pepper sauce
1 bay leaf
1/2 pound smoked or roasted turkey, cut into
    medium dice
1/2 cup light cream or half-and-half
1 large egg yolk, lightly beaten
2 tablespoons grated imported Parmesan cheese

To make the crust, in a medium bowl, combine the flour and salt. Using a pastry blender or 2 knives, cut in the shortening and butter until the mixture resembles coarse crumbs. Add the ice water, 1 tablespoon at a time, tossing until the dough clings together in a rough mass. Turn out onto a lightly floured surface and press the dough together into a flat round. Wrap in waxed paper and refrigerate for at least 30 minutes or up to 24 hours.

Preheat the oven to 400 degrees F.

To make the filling, in a deep skillet over medium heat, melt the butter. Add the leeks, bell pepper and celery and sauté slowly, stirring, 20 to 25 minutes, or until the vegetables are tender. Meanwhile, in another medium skillet, combine the calabaza with 1/2 cup of the stock. Cover and cook over low heat, 15 to 20 minutes, or until the squash is just tender.

When the leek mixture is tender, sprinkle with the flour and stir well. Over medium heat, slowly add the remaining 1 1/2 cups

warm stock, stirring constantly. Continue to cook, stirring frequently, for 5 to 10 minutes longer, or until the mixture turns smooth and fairly thick. Add the calabaza and its cooking liquid, the thyme, sage, hot-pepper sauce and bay leaf and mix well. Simmer gently for 5 minutes. Add the turkey, mix well and heat through. Stir in the cream.

Divide the turkey filling evenly among 4 small baking dishes, each approximately 4 1/2 inches in diameter and 2 1/2 inches deep. On a lightly floured board, roll out the dough 1/2 inch thick. Cut into rounds to fit the tops of the baking dishes. Transfer a dough round to each dish, crimping it against the edges. Prick each crust in a few places with a fork. Brush with the egg yolk and sprinkle with the Parmesan.

Bake in the oven for 20 to 30 minutes, or until the tops are golden brown and the filling is bubbly. Serve piping hot. *Serves 4*

Tandoori Lamb and Zucchini Kabobs with Zucchini Raita

*This adaptation of a classic Indian dish works because the marinade
makes the lamb and zucchini hot and the yogurt sauce cools them down. The kabobs
are best cooked on a grill, but will do fine under the broiler.*

1 1/4 pounds boneless leg of lamb, cut into
    1 1/2-inch cubes
1/2 cup plain yogurt
One 3-inch piece fresh ginger, peeled and coarsely
    chopped
4 cloves garlic
1 teaspoon cumin seeds
1 teaspoon coriander seeds
1/2 teaspoon yellow or black mustard seeds
2 teaspoons red pepper flakes
1 teaspoon salt

1 large red bell pepper, cored and seeded
1 small red onion
2 medium green or golden zucchini
4 mushrooms, stemmed

Raita:
1/2 cup plain yogurt
1/4 cup coarsely grated zucchini
1/4 cup finely diced ripe tomato
1 tablespoon finely chopped fresh cilantro
1 tablespoon yellow mustard seeds

Place the lamb cubes in a glass bowl or a nonreactive container. In a blender, combine the yogurt, ginger, garlic, cumin, coriander and mustard seeds and pepper flakes. Purée until smooth. Stir in the salt, then pour the mixture over the lamb, turning the cubes to coat evenly. Cover and marinate overnight in the refrigerator.

The next day, prepare a fire in a charcoal grill, or preheat the broiler. Cut the bell pepper and onion into chunks the size of the lamb cubes. Trim the ends of the zucchini. Cut into slices 1/2 inch thick. Remove the lamb from the marinade and thread onto 4 metal skewers alternately with the onion, bell pepper and zucchini pieces. Place a mushroom cap at the end of each skewer. Grill or broil, turning to brown all sides, until the meat is done, 6 to 10 minutes depending on how well-done you like your lamb.

While the kabobs are cooking, make the raita. In a small bowl, combine the yogurt, zucchini, tomato and cilantro and mix well. Place the mustard seeds in a nonstick skillet over high heat and toast a minute or so until they pop, then stir them into the yogurt mixture. Serve the kabobs with the raita on the side. *Serves 4*

Acorn Squash Stuffed with Chorizo and Corn

This spicy one-dish meal brings together the best of the Southwest pantry—
squash, chorizo sausage, corn, rice and green olives. Sweet dumpling or roly poly squash
can fill in for the acorn. Kielbasa or Italian sausage can replace the chorizo.

4 acorn squash, each approximately 4 inches in
    diameter
1/2 pound Mexican chorizo sausages, casings removed
1 cup cooked basmati or long-grain white rice
1 cup fresh corn kernels (or frozen corn, thawed)
4 green onions, green parts only, coarsely chopped

1/2 cup coarsely chopped pimento-stuffed
    green olives
1/4 cup chicken stock, preferably homemade,
    or water
Salt and freshly ground black pepper, to taste
3/4 cup grated Monterey Jack cheese

Preheat the oven to 350 degrees F.

Cut a thin slice off the bottom of each squash, removing just enough so that it will stand upright. Then slice off the tops (no more than 1 inch thick) and scrape out the seeds. Set aside.

In a small skillet, crumble the sausage and place over medium heat. Fry, stirring frequently, approximately 10 minutes, or until browned. Transfer to a mixing bowl, along with any fat in the pan. Add the rice, corn, green onions, olives and stock. Mix well, then

season with salt and pepper. Stir in 1/2 cup of the cheese.

Pack the mixture into the cavities of the prepared squash. Set the filled squash upright in a 9-by-13-inch baking dish and add hot water to the dish to a depth of approximately 1 inch. Cover loosely. Bake in the oven for 30 minutes. Uncover the dish and sprinkle the remaining 1/4 cup cheese evenly over the filled squash. Bake for 10 to 15 minutes longer, or until the squash can be easily pierced with a fork and the cheese is melted. *Serves 4*

Spaghetti Squash Brunch Bake with Cheddar

Like an American frittata, this quick and easy dish can be a brunch or lunch main course.
The cheese can be changed—try mozzarella, imported Parmesan or even chèvre—and zucchini
or other summer squash can be substituted for the spaghetti squash.

1 tablespoon unsalted butter, melted
1 small red bell pepper, cored, seeded and cut into
*   1/4-inch dice*
2 fresh jalapeño chilies, seeded and cut into
*   1/8-inch dice*
2 cups packed cooked spaghetti squash
1 cup fresh corn kernels (or frozen corn, thawed)

1 teaspoon dried oregano, preferably Mexican,
*   crumbled*
Salt and freshly ground black pepper, to taste
6 large eggs
1/2 cup heavy cream
2 cups grated extra-sharp Cheddar cheese

Preheat the oven to 350 degrees F.

In a small skillet over medium heat, melt the butter. Add the bell pepper and jalapeños and sauté approximately 5 minutes, or until tender. Transfer to a medium bowl. Add the squash, corn and oregano to the bowl. Mix well and season with salt and pepper.

In a second bowl, beat the eggs with the cream until blended. Pour over the squash mixture and add 1 1/2 cups of the cheese. Mix well. Pour into a 9-inch-square glass baking dish. Sprinkle the remaining 1/2 cup cheese evenly over the top.

Bake in the oven 30 to 40 minutes, or until set. Let cool slightly, then serve. *Serves 4 to 6*

Zucchini and Cheese Enchiladas with Tomatillo Sauce

*Enchiladas taste best to me when stuffed with something lighter than
the conventional chicken or beef. Grated zucchini retains its texture in slow baking,
while the pungent tomatillos and poblanos in the sauce ignite the flavor.*

4 large fresh poblano chilies (approximately 3/4 pound)
1 1/2 pounds tomatillos, husks intact
1 teaspoon dried oregano, preferably Mexican,
    crumbled
1/2 cup chicken stock, preferably homemade,
    or water

3 cups loosely packed, coarsely grated green zucchini
    (approximately 3 medium)
1 medium yellow onion, finely diced
4 cups grated Monterey Jack cheese with hot peppers
1 cup corn oil
12 corn tortillas

Preheat the broiler.

Arrange the poblanos on a heavy baking sheet and place under the broiler. Roast, turning to expose all surfaces until they are evenly charred and blistered on all sides. (The roasting time depends on the thickness of the peppers and the intensity of the broiler; figure on 10 to 15 minutes.) Let cool, then slip off the skins, core and remove the seeds. Pat dry with paper towels and place in a blender or in a food processor fitted with the metal blade.

Place the tomatillos, in their husks, in a dry heavy skillet over medium heat. Cover the pan and cook approximately 10 minutes, or until the tomatillos have patches of brown and are fairly soft but not mushy. Let cool slightly, then remove and discard the husks. Transfer to the blender or processor holding the poblanos. Add the oregano and chicken stock and purée until smooth.

Ladle 1 cup of the sauce into the bottom of an ungreased 9-by-13-inch glass baking dish and set aside. Transfer the remaining sauce to a shallow dish and keep ready near the stove.

Adjust the oven temperature to 350 degrees F.

In a medium bowl, combine the zucchini, onion and 2 cups of the cheese. Mix well and place next to the stove.

In a medium skillet over medium heat, warm the oil. Using tongs, dip each tortilla into the hot oil approximately 30 seconds to soften; drain off excess oil into the pan. Then dredge each tortilla in the tomatillo sauce to coat lightly; very little will cling to it. Lay the tortilla in the sauce-lined baking dish and spoon approximately 1/4 cup of the zucchini mixture down the center. Roll up into a cylinder and position seam side down in the dish. Repeat with the remaining tortillas, packing each filled tortilla tightly against the next. Distribute the remaining tomatillo sauce evenly over the enchiladas, coating the ends particularly well. Sprinkle the remaining 2 cups cheese evenly over the top.

Bake in the oven for 30 minutes, or until the tortillas are soft, the cheese is melted and blended with the filling and the edges are crisp. Serve hot. *Serves 4 to 6*

Hokkaido Risotto with Prosciutto and Parmesan

*With its subtle flavor, squash is particularly suited to a creamy risotto. Just cook
it separately or it will turn to mush. Pumpkin or butternut squash can fill in for the Hokkaido.*

1 small Hokkaido squash (approximately 3 pounds)
8 tablespoons (1 stick) unsalted butter
Approximately 7 cups rich duck, turkey or chicken
    stock, preferably homemade
4 slender leeks, white parts only, carefully washed
    and finely chopped

1 1/2 cups Arborio rice
2 ounces thinly sliced prosciutto, cut into
    julienne strips
Dash of red pepper flakes
1/2 cup grated Parmigiano-Reggiano cheese
Salt and freshly ground white pepper, to taste

Using a sharp knife, cut the squash in half. Scrape out the seeds and peel away the skin. Cut the flesh into 1/2-inch dice.

In a large skillet over medium heat, melt 3 tablespoons of the butter. Add the squash and sauté for approximately 15 minutes, or until just tender but still firm. Set aside in a warm spot while cooking the risotto.

In a large saucepan, bring the stock to a simmer, then select a wide, deep pan and place over medium heat. Melt the remaining 5 tablespoons of butter in the pan, add the leeks and sauté for approximately 15 minutes, or until soft. Stir in the rice and cook, stirring,

until all the grains are well coated with the butter. Stir in 1 cup of the hot stock and cook, stirring, until all the liquid is absorbed. Continue adding stock, 1 cup at a time, and stir constantly until each addition is absorbed. After 6 cups have been absorbed, stir in the reserved squash, the prosciutto and pepper flakes. Continue cooking, adding the remaining stock as needed, until the rice is al dente, with no starchy raw taste, and the mixture is slightly soupy but still appealing. Stir in half of the cheese and season with salt and pepper.

Ladle into serving dishes. Pass the remaining cheese. *Serves 3 or 4*

Pumpkin and Shiitake Ravioli with Sage Butter

Ravioli are surprisingly easy to make if you start with prepared
pasta dough or even wonton wrappers. Canned pumpkin makes the firmest filling,
or you can use a denser, drier squash like buttercup.

Ravioli:

3 tablespoons unsalted butter
1 tablespoon extra virgin olive oil
16 shiitake mushrooms, stemmed and finely diced
2 shallots, finely chopped
2 cloves garlic, minced
1 teaspoon tamari or soy sauce
1 teaspoon minced fresh sage
1 1/2 cups canned pumpkin purée or very dry
  buttercup purée

3/4 cup grated Parmigiano-Reggiano cheese, plus
  additional cheese, for garnish
1/2 cup fine dried bread crumbs
Salt and freshly ground black pepper, to taste
2 large egg whites
4 sheets fresh lasagna dough, each 16 by 24 inches,
  or 24 wonton wrappers

Sage Butter:

8 tablespoons (1 stick) unsalted butter
16 fresh sage leaves, julienned

To make the ravioli, in a large skillet over medium heat, melt the butter with the oil. Add the mushrooms, shallots and garlic and sauté for approximately 10 minutes, or until tender. Stir in the tamari and sage, raise the heat to high and cook, stirring constantly, 3 to 5 minutes, or until the mushrooms start to darken. Transfer to a mixing bowl. Let cool slightly, then stir in the pumpkin, cheese and bread crumbs. Season with salt and pepper. Blend in 1 egg white and set aside.

Cut the pasta dough into twenty-four 4-inch squares. Spoon a tablespoon of filling slightly off the center of each square. Brush the edges with egg white, fold over to form a triangle and press edges together tightly to seal. Wrap in plastic film and refrigerate for

at least 1 hour before cooking. (They can be stored overnight.)

Bring a large pot filled with water to a rolling boil. Add salt to taste and return to the boil. Gently lower the ravioli into the pot and cook until just al dente. This will take approximately 3 to 4 minutes, depending on the pasta. They are done when they float to the surface. Using a slotted spoon, lift out the ravioli and drain well, then arrange on a warmed serving platter.

To make the sage butter, melt the butter in a small saucepan over medium heat. Add the sage and cook for 2 to 3 minutes to infuse the butter. Drizzle the sage butter over the ravioli. Serve with additional cheese and black pepper. *Serves 4 to 6*

Penne in Spicy Tomato Sauce with Lita and Rosemary Meatballs

*Squash is usually just sliced or diced and added to a pasta sauce, but here it is
added to the meatball mixture with particularly juicy results. The sweet lita, an Italian variety,
is especially compatible with penne. Zucchini or yellow squash can be substituted.*

1/4 cup extra virgin olive oil
1 large yellow onion, finely diced
5 large cloves garlic, finely chopped
One 26-ounce can crushed tomatoes
3 tablespoons tomato paste
1 teaspoon dried oregano, crumbled
1 teaspoon dried basil
1 teaspoon red pepper flakes
1 bay leaf
1/2 cup coarsely chopped, pitted Kalamata olives

2 teaspoons salt, plus salt and freshly ground
    black pepper, to taste
1 pound ground round of beef
1 cup firmly packed, coarsely grated lita squash
    (approximately 3 small lita)
1 large egg
1 tablespoon finely chopped fresh rosemary
1/4 cup grated imported Parmesan cheese, plus
    additional grated Parmesan, for serving
3/4 pound dried penne

In a large, deep skillet or saucepan over medium heat, warm the oil. Add the onion and
sauté for approximately 5 minutes, or until
translucent. Add two-thirds of the garlic and
sauté for 2 minutes. Stir in the tomatoes,
tomato paste, oregano, basil, pepper flakes,
bay leaf and olives. Reduce the heat to low
and cook, stirring occasionally, for 30 minutes,
or until slightly thickened. Season with salt
and pepper and keep warm.

While the sauce is cooking, make the
meatballs. Break up the ground round into a
bowl. Add the squash, egg, rosemary, the 1/4
cup Parmesan and the remaining chopped
garlic. Mix until all the ingredients are thor-

oughly incorporated. Season with 1 teaspoon
of the salt and 3 or 4 grinds of black pepper.
Shape into 1 1/2-inch balls. Slip the balls into
the hot sauce, cover and simmer for 15 minutes, or until cooked through.

Meanwhile, bring a large pot filled with
water to a rolling boil. Add the remaining 1
teaspoon salt and bring back to a boil. Stir in
the penne and cook until al dente. This will
take approximately 11 minutes, depending on
the brand. Drain well and transfer to a large
serving bowl.

Pour the sauce and the meatballs over the
pasta and serve immediately. Pass the remaining Parmesan cheese. *Serves 4*

Grilled Mahi-Mahi on Rapid Ratatouille

Ratatouille is usually served solo after many hours of stewing, but I think this version makes a healthful and appealing partner for grilled fish. Salmon, tuna or swordfish steaks work as well as the mahi-mahi.

5 tablespoons extra virgin olive oil
1 small eggplant (approximately 1/2 pound), peeled and cut into 1/2-inch dice
1 medium Spanish onion, peeled and cut into 1/2-inch dice
1 large red bell pepper, cored, seeded and cut into 1/2-inch dice
3 large cloves garlic, finely chopped
4 small, slender green or golden zucchini, trimmed and cut into 1/2-inch dice
4 large, very ripe tomatoes, seeded and cut into 1/2-inch dice
1 tablespoon tomato paste
Dash of red pepper flakes
Salt and freshly ground black pepper, to taste
4 mahi-mahi fillets (approximately 5 ounces each)
1/4 cup chopped fresh basil, plus basil sprigs, for garnish
1/4 cup chopped, pitted Niçoise or Kalamata olives

In a large, deep skillet over medium-high heat, warm 3 tablespoons of the oil. Add the eggplant and sauté, stirring constantly, approximately 10 minutes, or until just soft. Transfer to a bowl and set aside.

Add 1 tablespoon of the remaining oil to the same skillet over medium heat. Add the onion, bell pepper and garlic and sauté approximately 10 minutes, or until tender. Return the eggplant to the skillet and add the zucchini, tomatoes, tomato paste and pepper flakes. Mix well, then cover and simmer gently for approximately 20 minutes, or until the vegetables are soft. Uncover and cook approxi-mately 5 to 10 minutes longer, or until the ratatouille mixture is fairly dry. Season with salt and pepper, re-cover and set aside off the heat.

Preheat a stove-top grill, or prepare a fire in a charcoal grill. Rub the fish fillets on both sides with the remaining 1 tablespoon oil. Season both sides with plenty of salt and pep-per. Grill the fish, turning once for approxi-mately 3 minutes on a side, or until just tender.

Stir the chopped basil and olives into the ratatouille. Spoon onto 4 individual plates and top each portion with a fish fillet. Garnish with basil sprigs and serve at once. *Serves 4*

SWEETS

My mother may have been negligent when it came to providing fresh vegetables, but somehow she managed to put dessert on the table every single night. Yet years of sugary cakes and cookies left me with a yearning for more savory foods. That's why I like squash as a dessert ingredient. I just think of it as a vegetable and treat it like a fruit.

Winter squash are so sweet on their own that they can fit into anything from a creamy cheesecake with cranberries to a sleek pie with a crunchy praline crust. They are most versatile when baked and puréed and then added to anything from sweet bread dough to aromatic Indian pudding. Butternut squash gives crème brûlée a little fiber to go with the richness and doubles the nutrients in a dense banana bread.

Even zucchini is as good as a carrot to a baker. It keeps bar cookies with black walnuts moist and peanut-oatmeal drop cookies chewy.

Because squash is so inexpensive, a dessert baker can indulge in pricier accoutrements, from black walnuts to crystallized ginger to sun-dried cranberries. All the "sweet" spices—ginger, cinnamon, nutmeg, allspice, mace, cloves—are particularly compatible with just about any squash, from acorn to zucchini. And because squash is so low in calories and high in nutrients, you can add sugar to your dessert and eat it, too.

Cranberry and Pumpkin Cheesecake in Gingersnap Crust

Two fall favorites come together in a rich and creamy dessert
that's easy to make. Just leave enough time to chill the cheesecake overnight.

1 1/2 cups finely crushed gingersnaps
6 tablespoons unsalted butter, melted
1 cup fresh or frozen cranberries
1/2 cup plus 2/3 cup granulated sugar
1/4 cup water
24 ounces cream cheese, softened

5 large eggs
2/3 cup sour cream
1 cup cooked, puréed pumpkin, fresh or canned, or
    other winter squash
1 teaspoon ground ginger

In a bowl, combine the gingersnap crumbs and the melted butter and mix well. Press the mixture onto the bottom and partway up the sides of a 10-inch springform pan. Cover the outside bottom and sides of the pan with aluminum foil. Refrigerate the crust while preparing the filling.

Preheat the oven to 325 degrees F.

In a small, heavy saucepan, combine the cranberries, the 1/2 cup sugar and the water and bring to a boil. Cook, stirring occasionally, for approximately 10 minutes, or until the berries pop and the mixture thickens. Let cool completely.

In a bowl and using an electric mixer set on medium speed, beat together the cream cheese and the 2/3 cup sugar until smooth. Add the eggs and sour cream and continue to beat until completely smooth. Transfer 1 cup of the mixture to a blender and set aside. Beat the pumpkin and ginger into the remaining batter until fully incorporated. Pour into the prepared crust. Add the cooled cranberries to the blender and purée until smooth. Pour this mixture over the pumpkin batter. Using a butter knife, swirl the cranberry mixture into the pumpkin batter.

Bake in the oven for approximately 1 hour, or until the filling is set around the edges. The center will still look jiggly. Turn off the oven, prop the door open and let the cheesecake cool down for 2 hours. Transfer to a rack and let cool completely. Refrigerate overnight before serving. *Serves 10 to 12*

Rum-Spiked Butternut Brûlée

Crème brûlée is irresistible in restaurants, but there's no reason to wait for dinner out.
The squash makes this version a little grainier than the French favorite, and the rum makes it headier.
Pumpkin or buttercup squash can be used instead of the butternut.

3 large egg yolks
1 cup heavy cream
1/2 cup firmly packed light brown sugar, plus
  additional light brown sugar, for topping
1 cup cooked, puréed butternut squash

2 tablespoons dark rum
1/2 teaspoon ground mace
1/4 teaspoon freshly grated nutmeg
Dash of salt

Preheat the oven to 350 degrees F.

In a bowl, combine the egg yolks, cream and the 1/2 cup brown sugar and beat until smooth. Stir in the squash, then mix in the rum, mace, nutmeg and salt. Stir until thoroughly mixed. Ladle into four 6-ounce ramekins or custard cups. Set the ramekins in a 9-by-13-inch glass baking dish. Pour hot water into the dish to reach halfway up the sides of the ramekins.

Bake in the oven for 30 to 40 minutes, or until set. Let cool completely, then cover and refrigerate at least 3 hours, or until very cold.

Just before serving, preheat the broiler. Sieve a thin, even layer of brown sugar over the top of each custard. Place under the broiler 3 to 4 inches from the heat source for 1 to 3 minutes, or just until the sugar caramelizes. Do not allow to burn. Serve at once. *Serves 4*

Banana and Butternut Squash Bread with Mace and Pecans

*Double your nutrients and your flavors: This bread is a hybrid of
classics using banana and squash. Vary the spices and add different nuts if you wish.*

1 cup granulated sugar
1/2 cup (1 stick) unsalted butter, softened
3/4 cup mashed very ripe banana
1 cup cooked and mashed butternut or other
    winter squash
2 large eggs
1 teaspoon vanilla extract
2 cups all-purpose flour

1 teaspoon baking soda
1/2 teaspoon baking powder
1/2 teaspoon salt
1 teaspoon ground mace
1/2 teaspoon freshly grated nutmeg
1 cup coarsely chopped pecans or walnuts
1/2 cup toasted, hulled pumpkin seeds (see p. 10),
    (optional)

Preheat the oven to 350 degrees F. Butter a
9-by-5-inch loaf pan, preferably glass.

In a large bowl and using an electric mixer
set on medium speed, beat together the sugar
and butter until light. Beat in the banana and
squash. Add the eggs, beating until smooth,
then mix in the vanilla.

In another bowl, stir together the flour,
baking soda, baking powder, salt, mace and

nutmeg. Add the flour mixture to the butter
mixture, beating until smooth. Stir in the nuts.
Spread the batter into the prepared pan and
sprinkle with the pumpkin seeds, if desired.

Bake in the oven for 50 to 60 minutes, or
until a toothpick inserted in the center comes
out clean. Let cool completely on a rack before
slicing. *Makes 1 loaf*

Buttercup Praline Pie

A cross between a pumpkin and a pecan pie, this rich dessert should please everyone at your holiday gathering. Pumpkin, butternut, Hokkaido, kabocha or Hubbard squash can also be used.

Crust:
1 cup plus 2 tablespoons unsifted all-purpose flour
1/2 teaspoon salt
1/3 cup chilled shortening
2 1/2 to 3 tablespoons ice water

Filling:
1 1/2 cups mashed, cooked buttercup squash
1/2 cup firmly packed dark brown sugar
1 teaspoon ground ginger
1/2 teaspoon ground cinnamon
1/2 teaspoon ground allspice

1/4 teaspoon grated nutmeg or ground mace
1/8 teaspoon ground cloves
1/8 teaspoon salt
2 large eggs, lightly beaten
1 cup light cream or half-and-half

Topping:
1/3 cup all-purpose flour
1/3 cup firmly packed dark brown sugar
3 tablespoons unsalted butter, chilled, cut
   into chunks
1/3 cup finely chopped pecans

To prepare the pie crust, in a large mixing bowl, combine the flour and the salt. Using a pastry blender or two knives, cut in the shortening until the mixture resembles coarse meal. Sprinkle a little of the cold water over the top. Toss with a fork until the mixture clings together, then add more water. The dough should not be sticky. Continue adding water and tossing with a fork until the dough comes together in a rough ball. Turn out onto a lightly floured surface and knead lightly with the heel of your hands until the dough makes a smooth ball. Wrap dough in wax paper and chill at least 30 minutes or overnight in the refrigerator.

Preheat the oven to 375 degrees F.

Roll out the dough on a lightly floured surface into a circle approximately 12 inches in diameter. Carefully lift into a 9-inch pie plate. Press lightly into the pan and crimp the edges with a fork. Cover with plastic wrap and

refrigerate for 15 minutes. Bake for 20 minutes, or until lightly golden. Remove and let cool.

Raise the oven heat to 450 degrees F.

In a large bowl, combine the squash and brown sugar and mix until smooth. Stir in the spices and salt, mixing well. Then stir in the eggs and cream until fully combined. Pour into the cooled pie shell and set aside.

To make the topping, in a second, smaller bowl, stir together the flour and sugar. Using a pastry blender or 2 knives, cut in the butter until the mixture resembles chunky crumbs. Add the pecans and mix well. Crumble the mixture over the top of the pie filling.

Bake in the oven for 10 minutes. Lower the oven temperature to 350 degrees F. and bake for 35 to 40 minutes longer, or until a knife inserted in the center comes out clean. Let cool completely on a rack before serving.
Serves 8 to 10

Super Spicy Indian Pumpkin Pudding

*A quintessential American dessert gets color and kick from
pumpkin in this aromatic variation. Other winter squash such as butternut,
Hubbard, buttercup, Hokkaido or kabocha can be substituted.*

1/4 cup coarse-grind cornmeal
1 cup water
3 cups milk
1 teaspoon salt
1 cup cooked, puréed pumpkin, fresh or canned
1 large egg, lightly beaten
1/3 cup firmly packed dark brown sugar

1/2 cup molasses
1 tablespoon unsalted butter
1 teaspoon ground ginger
1/2 teaspoon ground allspice
1/8 teaspoon ground cloves
Vanilla ice cream (optional)

Preheat the oven to 325 degrees F. Lightly butter a 1 1/2-quart soufflé dish or casserole and set aside.

In a bowl, stir the cornmeal into the water, mixing well. Transfer to a heavy saucepan and stir in 2 cups of the milk and the salt, blending well. Bring to a boil, stirring constantly. Cook over medium-high heat, stirring, 10 minutes longer, or until thick and smooth. Remove

from the heat and stir in the pumpkin, egg, sugar, molasses, butter, ginger, allspice and cloves. Pour into the prepared dish.

Bake in the oven for 30 minutes. Remove from the oven and stir in the remaining 1 cup milk. Return to the oven and bake for 1 1/2 hours longer, until thick and bubbly. Serve warm. Accompany with vanilla ice cream, if desired. Serves 6

Pumpkin and Chutney Kolaches

*I discovered kolaches, the Czech version of Danish pastries, while living in Nebraska and
found the not-too-sweet dough takes quite well to pumpkin. The filling is usually apricots or poppy seeds,
but the chutney works better with this version. These are best eaten the same day they're baked.*

1/4-ounce envelope active dry yeast (scant 1 tablespoon)
4 to 4 1/2 cups all-purpose flour
1/2 cup (1 stick) unsalted butter, softened, plus 2
    tablespoons unsalted butter, melted
1/2 cup milk

3/4 cup granulated sugar
1/4 teaspoon salt
2 large egg yolks, at room temperature
1/2 cup cooked, puréed pumpkin, fresh or canned
1/2 cup mango chutney, finely chopped

In a large bowl, combine the yeast and 1 1/2 cups of the flour. In a small saucepan, combine the 1/2 cup butter, milk, sugar and salt and heat until just scalded (approximately 115 degrees F.). Pour the warm mixture into the yeast mixture. Add the egg yolks and pumpkin and, using an electric mixer set on low speed, beat 1 minute. Increase the speed to high and beat 3 minutes longer. Using a wooden spoon, stir in enough of the remaining 2 1/2 to 3 cups flour to make a soft dough.

Turn out the dough onto a lightly floured surface and knead for 8 to 10 minutes, or until soft and elastic. Shape into a ball and place in a well-buttered bowl. Turn the dough to coat the surfaces with butter. Cover the bowl with a kitchen towel and let the dough rise in a warm, draft-free spot for approximately 1 1/2 hours, or until doubled in bulk.

Punch down the dough. Tear off pieces approximately the size of golf balls and form into smooth balls. Arrange 3 inches apart on buttered baking sheets. Cover with a kitchen towel and let the dough rise again for approximately 45 minutes, or until doubled in bulk.

Meanwhile, preheat the oven to 400 degrees F.

When the rolls have risen, flatten them slightly and make a small depression in the center of each. Spoon approximately 1 teaspoon of the mango chutney in each depression. Brush the rolls lightly with melted butter.

Bake 12 to 15 minutes, until golden brown. Serve warm. *Makes approximately 2 dozen kolaches*

Zucchini, Peanut and Oatmeal Cookies

Chewy and healthful, these moist drop cookies are quick to make and easy to adapt. Try chocolate chips, raisins or sun-dried cherries in place of the dates. Yellow squash or even winter squash such as butternut can be used instead of zucchini. These cookies are not good keepers; plan on eating them within a day of baking.

1/2 cup (1 stick) unsalted butter, softened
1/2 cup chunky peanut butter
3/4 cup firmly packed light brown sugar
3/4 cup granulated sugar
2 large eggs
1 teaspoon vanilla extract

1 1/2 cups all-purpose flour
1/2 teaspoon baking soda
1/2 teaspoon salt
1 cup old-fashioned oats
1 1/2 cups loosely packed, coarsely grated green zucchini
1 cup finely chopped, pitted dates

Preheat the oven to 375 degrees F.

In a large bowl and using an electric mixer set on medium speed, beat together the butter, peanut butter and sugars until smooth. Beat in the eggs and vanilla.

In another bowl, stir together the flour, baking soda and salt. Stir the flour mixture into the egg mixture until fully incorporated. Then stir in the oats, zucchini and dates, mixing well. Drop the dough by rounded teaspoonfuls onto ungreased baking sheets, spacing them 2 inches apart.

Bake in the oven for approximately 15 minutes, or until lightly browned. Let cool slightly on the baking sheets, then transfer to racks to cool completely. *Makes approximately 4 dozen cookies*

Zucchini and Black Walnut Bars with Cream Cheese Frosting

Moist and chewy, these easy bar cookies owe their texture to zucchini but their haunting flavor to black walnuts. If you can't find black walnuts in a health-food store or specialty market, try pecans or regular walnuts. Grated winter squash can fill in for the zucchini.

1 1/2 cups loosely packed, coarsely grated
    green zucchini
2 large eggs
1 cup granulated sugar
3/4 cup corn oil
1 1/2 teaspoons vanilla extract
1 1/2 cups all-purpose flour
1 teaspoon baking powder
1/2 teaspoon salt
1 teaspoon ground cinnamon

1/2 teaspoon ground allspice
1/4 teaspoon freshly grated nutmeg
1 cup black walnuts, coarsely chopped

Frosting:
3 ounces cream cheese, softened
Approximately 2 cups sifted confectioners' sugar
2 tablespoons milk
Dash of salt
1 teaspoon vanilla extract

Preheat the oven to 350 degrees F. Lightly oil a 9-by-13-inch baking dish and set aside.

Place the zucchini in a colander to drain, pressing out as much moisture as possible and set aside.

In a large bowl and using a wooden spoon, beat together the eggs, sugar and oil until well blended. Beat in the vanilla.

In another, smaller bowl, stir together the flour, baking powder, salt, cinnamon, allspice and nutmeg. Stir the flour mixture into the egg mixture, blending well. Add the zucchini and walnuts and mix well. Spread into the prepared baking dish.

Bake in the oven for 30 minutes, or until the top springs back when lightly pressed. Let cool completely on a rack.

To make the frosting, in a bowl and using a wooden spoon or an electric mixer set on medium speed, beat the cream cheese until smooth. Then beat in the sugar and milk, adding just enough sugar to form a spreadable consistency. Stir in the salt and vanilla until well mixed.

Spread the frosting over the cooled cake. Using a sharp knife, carefully cut into bars, cutting lengthwise into 4 strips and then across into 8 strips. *Makes 32 bars*

Kabocha, Cranberry and Ginger Tart in an Almond Crust

*Squash is normally puréed for desserts, but it has a superb texture when
simply grated and added raw to this tart filling. Serve with whipped cream, if you like.
Calabaza or butternut squash can be used instead of kabocha.*

Pastry:
1 1/3 cups all-purpose flour
1/3 cup ground almonds
1/4 cup granulated sugar
1/2 teaspoon salt
1/2 cup (1 stick) unsalted butter, chilled,
    cut into chunks
1 large egg yolk
1 tablespoon freshly squeezed lemon juice
2 to 3 tablespoons ice water

Filling:
2 cups coarsely grated kabocha squash
1/4 cup finely diced crystallized ginger
1/2 cup sun-dried cranberries
3 large eggs
1/2 cup honey
1/2 cup heavy cream
4 tablespoons (1/2 stick) unsalted butter, melted
1/4 teaspoon salt

To make the pastry, in a bowl, combine the flour, almonds, sugar and salt. Using a pastry blender or 2 knives, cut in the butter until the mixture resembles coarse meal. In a small bowl, beat together the egg yolk and lemon juice. Sprinkle over the dough and toss together with a fork. Add the ice water, a tablespoon at a time, and continue tossing until the ingredients start to cling together in a rough mass. Turn out onto a lightly floured surface and press a couple of times with the heel of your hand to combine the ingredients. Shape into a flat round, wrap in waxed paper and refrigerate for at least 1 hour or up to 24 hours.

Position a rack in the bottom third of the oven. Preheat the oven to 425 degrees F.

Place the dough round on another sheet of waxed paper. Roll out the dough into a round to fit the bottom and up the sides of a 9-inch tart pan with a removable bottom. Transfer the round to the pan, easing it in gently. Trim off the excess crust even with the pan rim. Line with aluminum foil and fill with pie weights or dried beans. Bake in the bottom third of the oven for 10 minutes. Remove the weights and foil and bake for 5 minutes longer; the crust should be just starting to brown at the rim. Let cool completely on a rack.

Reduce the oven temperature to 350 degrees F.

To make the filling, in a small bowl, stir together the squash, ginger and cranberries. In another small bowl, lightly beat the eggs with the honey, cream and butter. Add to the squash mixture, along with the salt. Mix until fully combined. Pour into the prepared crust.

Bake in the oven for 45 to 50 minutes, or until set. Let cool completely before serving.
Serves 8 to 10

METRIC CONVERSIONS

Liquid Weights

U.S. Measurements	Metric Equivalents
1/4 teaspoon	1.23 ml
1/2 teaspoon	2.5 ml
3/4 teaspoon	3.7 ml
1 teaspoon	5 ml
1 dessertspoon	10 ml
1 tablespoon (3 teaspoons)	15 ml
2 tablespoons (1 ounce)	30 ml
1/4 cup	60 ml
1/3 cup	80 ml
1/2 cup	120 ml
2/3 cup	160 ml
3/4 cup	180 ml
1 cup (8 ounces)	240 ml
2 cups (1 pint)	480 ml
3 cups	720 ml
4 cups (1 quart)	1 litre
4 quarts (1 gallon)	3 3/4 litres

Dry Weights

U.S. Measurements	Metric Equivalents
1/4 ounce	7 grams
1/3 ounce	10 grams
1/2 ounce	14 grams
1 ounce	28 grams
1 1/2 ounces	42 grams
1 3/4 ounces	50 grams
2 ounces	57 grams
3 ounces	85 grams
3 1/2 ounces	100 grams
4 ounces (1/4 pound)	114 grams
6 ounces	170 grams
8 ounces (1/2 pound)	227 grams
9 ounces	250 grams
16 ounces (1 pound)	464 grams

Temperatures

Fahrenheit	Celsius (Centigrade)
32°F (water freezes)	0°C
200°F	95°C
212°F (water boils)	100°C
250°F	120°C
275°F	135°C
300°F (slow oven)	150°C
325°F	160°C
350°F (moderate oven)	175°C
375°F	190°C
400°F (hot oven)	205°C
425°F	220°C
450°F (very hot oven)	230°C
475°F	245°C
500°F (extremely hot oven)	260°C

Length

U.S. Measurements	Metric Equivalents
1/8 inch	3 mm
1/4 inch	6 mm
3/8 inch	1 cm
1/2 inch	1.2 cm
3/4 inch	2 cm
1 inch	2.5 cm
1 1/4 inches	3.1 cm
1 1/2 inches	3.7 cm
2 inches	5 cm
3 inches	7.5 cm
4 inches	10 cm
5 inches	12.5 cm

Approximate Equivalents

1 kilo is slightly more than 2 pounds
1 litre is slightly more than 1 quart
1 meter is slightly over 3 feet
1 centimeter is approximately 3/8 inch

INDEX

Acorn Squash Stuffed with Chorizo and Corn, 64

Baby Squash Stuffed with Chives and Maytag Blue, 27

Bread, Banana and Butternut Squash, with Mace and Pecans, 79

Buttercup Sauté, Fast and Spicy Shredded, 54

Butternut Bisque, Creamy, 21

Butternut Brûlée, Rum-Spiked, 78

Chayote, Steamed, in Cilantro Cream, 49

Cheesecake, Cranberry and Pumpkin, in Gingersnap Crust, 77

Chowder, Summer Squash and Corn, with Smoked Trout, 22

Cookies, Zucchini, Peanut and Oatmeal, 88

Crostini, Lita, with Goat Cheese and Tomato Tapenade, 33

Delicata, Roasted, with Red Chili and Lime Butter, 50

Enchiladas, Zucchini and Cheese, with Tomatillo Sauce, 66

Fritters, Bajan Calabaza and Cod, 29

Grilled Summer Squash with Herbs, 47

Kolaches, Pumpkin and Chutney, 87

Lamb and Zucchini Kabobs, Tandoori, with Zucchini Raita, 63

Madeleines, Savory Zucchini and Cheese, 32

Mahi-Mahi, Grilled, on Rapid Ratatouille, 72

Muffins, Kabocha Corn, with Sweet Onions, 34

Penne in Spicy Tomato Sauce with Lita and Rosemary Meatballs, 71

Pie, Buttercup Praline, 82

Pot Pies with Parmesan Crust, Turkey and Calabaza, 60

Pudding, Super Spicy Indian Pumpkin, 85

Ravioli, Pumpkin and Shiitake, with Sage Butter, 68

Risotto, Hokkaido, with Prosciutto and Parmesan, 67

Salad, Roasted Green and Golden Zucchini, with Peppers and Mozzarella, 37

Shaker Yellow Velvet, 49

Soufflé, Southern Crookneck, 42

Soup, Caramelized Pumpkin and Onion, with Smoked Turkey, 25

Soup, Chilled Curried Sunburst, with Tomato and Avocado, 21

Spaghetti Squash Brunch Bake with Cheddar, 65

Spaghetti Squash Primavera, 58

Squash Blossoms, Deep-Fried, 26

Sunburst Rémoulade, 50

Sunburst Squash Stuffed with Spinach and Gruyère, 53

Tart, Kabocha, Cranberry and Ginger, in an Almond Crust, 91

Turban Squash Purée, Thyme-Scented, 55

Zucchini and Black Walnut Bars with Cream Cheese Frosting, 89

Zucchini, Oven-Fried, in a Crunchy Parmesan Crust, 41

Zucchini, Tomato and Basil Gratin, 44

Zucchini Triangles, Pesto and, 35